Marjorie Marianela Asanza Castillo
Jordy Duval Cepeda Rico
Daniela A. Chuez Zamora

Production Of Californian Red Earthworms

Marjorie Marianela Asanza Castillo
Jordy Duval Cepeda Rico
Daniela A. Chuez Zamora

Production Of Californian Red Earthworms

ScienciaScripts

Imprint

Cover image: www.ingimage.com

This book is a translation from the original published under ISBN 978-620-2-11396-0.

Publisher:
Sciencia Scripts
is a trademark of
Dodo Books Indian Ocean Ltd. and OmniScriptum S.R.L publishing group

120 High Road, East Finchley, London, N2 9ED, United Kingdom
Str. Armeneasca 28/1, office 1, Chisinau MD-2012, Republic of Moldova, Europe
Printed at: see last page
ISBN: 978-620-6-22084-8

SUMMARY

The objective of this research was to evaluate the effect of three organic wastes for the production of *Eisenia foetida*, its lombrihumus, leachate and nutrient concentration; fed with three sources that were used as treatments: organic wastes (TI), bovine manure (T2) and plantain rachis (T3), with four replicates. A Randomised Complete Block Design was established. The variables studied were: number of young worms, adult worms without clitellum and adult worms with clitellum, eggs.

The results showed that there was a significant difference in the number of adult worms without clitellum. The concentration of nutrients in humus indicates that the primary macronutrients of the T3 obtained better results 1,8 (N) 0,57 (P) 1,60 (K) and the T1 in the secondary nutrients 2,13 (Ca) 0,59 (Mg), the pH was slightly alkaline. For the leachate analysis T1,T2,T3 corresponding to the element of N were presented in equal concentration 0,06, observing a difference in P in T1 of 0,04 and in K the T2 0,49 was higher than the other treatments, In the secondary elements analysed there is a minimum percentage of concentration in the different treatments. The humus production was better in T1 with 19.5

kg and for the leachate production T2 with 11 L.

Key words: Clitellum, cattle manure, leachate, Lombrihumus, plantain rachis, organic waste.

CHAPTER I

1.1. Introduction

Worm farming today is an alternative activity in agriculture, which is governed by rules similar to those used for the production of any domestic animal, requiring knowledge of the biology of the annelids and the technology for their breeding, feeding and reproduction. In the United States, earthworm farming dates back some 50 years, with the Californian red worm being the most commercial (Ferruzzi, 2009).

Vermiculture consists of the intensive cultivation of the red earthworm (*Eisenia foetida*) which consumes organic waste that is transformed and used as fertiliser for agricultural crops (Flores, 2010). The organic wastes thrown by the earthworm are known as humus, which is the highest state of decomposition of organic matter and is an excellent quality fertiliser.

The Californian red earthworm (*Eisenia foetida*), of the order annelid, class oligochaete and family lombrícidae, contributes to soil improvement by maintaining fertility due to the fact that it increases the availability of nitrogen, phosphorus and sulphur and inhibits the growth of fungi and bacteria that can affect the crop. The breeding of this earthworm is basically due to obtaining as a by-product the lombrihumus, which is digested and excreted organic matter that serves as a biofertiliser when in contact with the plant. *Eisenia foetida* is by nature a useful agent for recycling farm waste and is of special interest for organic production as it allows the transformation of waste into natural fertilisers that can be applied to the soil (Cando, 2010).

Vermiculture has aroused interest in countries such as the United States, Italy, Argentina, Cuba and France, which are processing large quantities of organic waste, generating fertilisers of excellent quality, which help to avoid the indiscriminate use of chemical fertilisers, making it a viable, efficient and environmentally friendly alternative, with low investment costs. The transformation of these organic wastes produces benefits by improving the physical, chemical and biological properties of soil permeability, cation exchange (Ferruzzi, 2009), and the physical and chemical qualities of lombrihumus favour increased plant growth and crop productivity (Dimas *et al.*, 2008). Lombrihumus is a stabilised material with a peat-like appearance, finely divided with high porosity, aeration, drainage and water holding capacity and a low C:N ratio (Domínguez *et al., 2010),* Humus, known in the world as "black gold", is an organic fertiliser free of chemicals, its use increases by 50% in any type of production, and can be used to recover land eroded by the indiscriminate use of chemical fertilisers, which are the cause of the loss of flora and fauna in the environment (Dalzell, 2011), as well as being used as a soil cover, organic fertiliser, or as an additive in chemical fertilisers (García *et al.,* 2005). Suppression of phytopathogenic soil fungi in organic waste composts has been demonstrated (Reyes *et al.*, 2007). The agronomic quality of lombrihumus places it as a material rich in organic matter (55.6 - 63.1 %), rich in major nutrients (1.87 - 2.13 % nitrogen and 2.06 - 2.52 % phosphorus), with a high value of cation exchange capacity (2.25 - 1,636 meq/100 g), higher than the values found in soils, which indicates that if this waste is used in agriculture it will provide an increase in soil fertility.

1.2. Objectives

1.2.1. General objective

> To evaluate the effect of three food substrates based on organic waste, bovine manure and banana rachis, for the production of *Eisenia foetida*, lombrihumus and leachate.

1.2.2. Specific objectives

> Determine the best substrate for *Eiseniafoetida* reproduction and growth.

> Evaluate the production of vermicompost and its leachate on each of the different substrates.

> To analyse the nutrient concentration levels in each of the different substrates.

CHAPTER II

2.1. Theoretical framework

2.1.1. Vermiculture Concept

Vermiculture is a biotechnology that uses a domesticated species of earthworm as a working tool, recycles all kinds of organic matter and produces two products: humus and worm meat (Ferruzzi, 2009).

The United States and Switzerland are the countries, which mainly practice vermiculture. In Latin America, we can mention: Ecuador, Argentina, Chile, Colombia, Costa Rica, Mexico, Nicaragua, Panama, Paraguay, Peru, Uruguay and Venezuela (Alastre, 2009).

2.1.2. History of vermiculture

The main earthworm-producing countries in Latin America are Chile, Brazil, Colombia, Argentina and Ecuador. These countries have large industrial farms of Californian red worm castings. The Philippines is one of the largest producers of worm meal for human consumption, as the absence of odour and taste makes it competitive with fish meal, both in quality and price (Esteve, 2007).

2.1.3. The Californian red worm

It is known as the Californian red earthworm because it is in this US state where its ecosystem properties were discovered and where the first farms were established and where it was bred intensively from the 1950s onwards in California.

According to (Rodríguez, F., 2010) they originate from Eurasia and are a species called "Hybrid Red", the most cultivated species in the world, due to its hardiness, tolerance to environmental factors, reproductive potential and capacity for crowding.

2.1.4. Care in vermiculture management

According to (De La Ossa, 2003), he states that care in vermiculture management is as follows:

a) The operator:

- Plastic gloves should be worn while handling manure, worms and humus and hands should be washed with soap and water at the end of the work.
- They should be vaccinated against tetanus (De La Ossa, 2003).

b) Cultivation:

- Feed should be provided to the worms whenever they require it, minimum once a week.
- Observe well the presence of cocoons and small or newly hatched worms, if they do

not increase, the protein content of the litter should be improved.

- It is necessary to harvest the adult worms to allow the growth and multiplication of the remaining litter.
- The crop should be checked periodically to ensure that the worms are still on the site, that the humidity and temperature are adequate and that no predators are present (De La Ossa, 2003).

2.1.5. Preparation of rearing beds

According to (Zenato, 2010), the suitable habitat for earthworms is the bed, in which they find all the basic requirements, which prevents them from escaping from underneath or from the sides.

The beds can be 1 m wide and 10 m long, with a height of 25 cm; the material to be used can be wood, bamboo cane, wood logs, concrete, bricks and/or any other non-rusting material. The orientation of the beds must be such as to allow all excess water to drain out, as water accumulated under the beds kills the earthworms. The beds should be constructed in the main wind direction and in exposure to the maximum amount of sunlight (Arledge, 2010).

2.1.6. Preparation of organic waste and worm castings mixture

The breeding beds are filled with the following layers; the first layer should be placed with a thickness of 5 to 7 centimetres of dry grass, the second layer of black soil, sprinkled along the bed with a thickness of 5 cm, then the third layer is placed the red Californian earthworm breeding foot along the bed, finally the layer of organic waste is placed on the layer of earthworms, these should be crushed to degrade faster (Cando, 2010).

2.1.7. Morphological and physiological characteristics

According to (Diaz, 2009), earthworms have a cylindrical and elongated body consisting of two concentric tubes: the body wall and the digestive tract, separated by the coelom and divided into segments called somites, with an anterior and posterior wall.
At birth the worms are white, after 5 or 6 days they turn pink and after 120 days they resemble adults due to their reddish colour and are ready to mate.

According to (Rodríguez, F., 2010) among the most important morphological characteristics of (*Eisenia foetida*) the following are mentioned:

- **Colour:** it has an intense reddish colour, which is why it is known as Californian Red.
- **Shape:** the body is a bilaterally symmetrical tube with a cylindrical shape.
- **Segments:** also called metameres, these are rings distributed throughout the body, usually comprising 80 to 175 rings.
- **Prostomium:** small dorsal protuberance beginning at the first segment, from which it is separated by a groove.
- **Peristomium:** this is the name given to the first segment, where the mouth is located; it does not have any quetae or bristles.

- **Quetas or bristles:** each segment, with the exception of the first one, has four pairs of quetas or bristles, provided with small muscles, whose function is locomotion.
- **Dorsal pores:** small openings located in the intersegmental grooves along the dorsal midline.
- **Nephridiopores:** paired excretory openings repeated in each body segment.
- **Seminal furrows:** located in segments 9 and 10, formed during copulation, they are transitory and store the spermatozoa received during copulation.
- **Clitellum: it** is the thickened region of the epidermis in segments 32 to 37 and its function is related to reproduction (Rodríguez, F., 2010).

2.1.8. Life cycle

Worms are incomplete hermaphrodites, they have both sexes complete, so they cannot self-fertilise and therefore copulation is necessary, which occurs every 7 to 10 days. As a result of this mating of two worms, two eggs or capsules (one from each worm) containing 2 to 21 worms will be produced.

Eisenia foetida begins its reproductive stage at 3 months, which continues throughout life. Mating takes place at night, above the soil surface, and usually lasts 30 minutes to 4 hours, producing a capsule every 7 to 10 days. After 14 to 21 days of incubation, approximately 4 to 20 daughter worms hatch.

The male genital tract consists of the testicles, which are sperm-secreting glands located very close to the mouth. The female genital tract receives the sperm and retains it until the moment of fertilisation, which is located posterior to the male genital tract. At the moment of coupling, they rotate in the opposite direction to each other, making contact between the male and female genitalia.

2.1.9. Taxonomic classification

According to (Dalzell, 2011) the Californian red worm has the following taxonomic characteristics:

Table 1. ***Taxonomic classification***

Kingdom:	Animal
Sub Kingdom:	Metazoa
Type:	Annelid
Phylum:	Protostomy
Class:	Annelid
Order:	Oligochaeta
Family:	Lumbricidae
Gender:	Eisenia
Species:	Foetida

Source: (Cando, 2010)

2.1.10. General aspects of production and its development

Location

In principle, the box can be placed anywhere, preferably in a place that is easily accessible for normal watering and feed distribution operations. From this perspective, garages, cellars, cellars, basements, attics, balconies and windows are advisable, provided that all these places are sufficiently ventilated and away from direct sources of heat or cold.

Temperature.

In order for the worms to develop well, the temperature has to be at a value of 25°C, however, in a range of not less than 15°C and not more than 30°C, they can still have a good performance, below this range the worms enter a period of dormancy decreasing their activity and exceeding these ranges, they stop producing capsules or coconuts which is where the new offspring come from to the populations themselves, since the temperature is fundamental for the fecundity of the coconuts (Flores-Pacheco, 2010).

Humidity.

Humidity shall be 70% to facilitate feed ingestion and sliding through the material. Moisture and nutrition are considered to be the most important factors for earthworms, as these animals do not have an adequate water conservation mechanism, although they require moisture in the body wall for respiration and lose a lot of water in their urine.

pH.

The earthworm accepts substrates with pH 5 to 8.4. Ranges outside this range are problematic for the earthworm. Substrate preparation should be done, if necessary, by aerobic fermentation.

Irrigation

Manual and sprinkler irrigation systems are used.

Aeration

It is important for the correct respiration and development of the worms, as inadequate respiration reduces food consumption, mating and reproduction due to compaction,

2.1.11. Vermiculture management

Feeding

According to Méndez (2015), he states that earthworms eat almost any putrefied organic substance and are very greedy for sugars, salts and cellulose. The finer the granules of the feed, the less difficulty they will have to ingest it and therefore the higher the humus production, so it is essential that the farmer crushes the feed before supplying it, to accelerate the degradation process and improve the texture.

For (Tineo, 2010) "the selection of the feed will depend largely on its availability at the site of the farm, this is one of the factors that determine the feasibility and location of the vermiculture".

The worms obtain their nutrition from the micro-organisms that grow in the organic matter, at the same time promoting a higher microbial activity in the waste produced, which is much more fragmented and microbially active than the original matter consumed, it is necessary that the quality of the food provided is the best (Peña, 2010).

- **Frequency and quantity of feed provided.** The best feeding system is free consumption, i.e. providing feed according to its demand, it is usual to provide feed 1 to 2 times per week in layers no bigger than 5 cm, keeping feeding records of the vermicompost (Peña, 2010).

- **Moisture requirements and frequency of watering.** The feed is prepared by soaking it before it is taken to the worm beds, soaking it if necessary, until it is completely wet and does not drain, this corresponds approximately to a range of 80 to 85 % humidity, the beds must also be soaked to conserve humidity, the leached liquids can be used as foliar fertilisers (Peña, 2010).

Diseases and pests

According to (Velásquez, 2003), the pests that attack the Californian earthworms in the different stages of development are:

- The ants
- Centipede
- Dung cockroaches
- Mites
- Snails

- Insects

- Birds
- Various rodents
- Snakes and lizards

Humus

Humus is classified according to the time of decomposition, into young and old, its transformation into fulvic acids and these in turn into humic acids where these molecules react to form humins, the latter two compounds have a beneficial effect on plant nutrition, due to their high efficiency in the exchange of cations.

The use of humus is of great importance as it helps to protect soils from erosion, and provides nutrients to plants and other plant organisms in the soil. Two of the components in the material are humic and fulvic acids, which are responsible for many of the improvements that humus exerts (Herrán, 2008). Humic substances increase the cation exchange capacity of soils by forming clay-humic complexes (Zabala 2003, 1993; Durán 2009; Herrán, 2008).

In relation to soil chemical properties, the application of lombrihumus improves soil fertility by increasing the content of plant-available nutrients (Flores *et al.*, 2009); increases soil organic C content (Durán *et al.*, 2013; Flores *et al.*, 2009); and modifies pH, electrical conductivity (EC) and cation exchange capacity (CEC) (Flores *et al.*, 2009).

Table 2. Humus type

	Young humus	**Old humus**
Characteristic -	• Bajadescomposición • Dark colouring Presence of acids fulvic acids and • humic acids • Preferred bysualta • of nutrientsits	- Altadescomposition - Dark colouring, between - violet and ochre - Presence of humic acids and - humins - Preferred by ability to quantityretain water in the soil and avoid erosion

Source: (Garcia, 2009).

Leachate

According to (Stanley, 2006) it is the liquid resulting from the decomposition of waste, originating from its own humidity, reaction, dragging or dissolution of a solvent or water when in contact with it.

Leachates are a by-product of the decomposition of matter (organic and inorganic), depending on the composition of the materials their degree of toxicity will be determined, with organic plant materials being less toxic than those of animal origin, as well as materials based on chemical products, plastics, metals, radioactive materials and biological materials (Del Pilar, 2013).

Leachate characteristics

According to (Castells, 2012) several disease suppressing effects of leachates are detailed:

- Inhibition of spore germination on diseased plants.
- Stopping the spread of the lesion on the surface of the plant.
- Competition with micro-organisms for food and nutrients.
- Predation of disease-causing micro-organisms.
- Elimination of antibiotic-producing organisms.
- Increase of the plant's health and, with it, its ability to defend itself against diseases.

CHAPTER III

3.1. Methodology

3.1.1. Geographical location of the trial

This research was carried out in the months of July 2018 to February 2019, located in the San Pablo II area of the Canton of Valencia in the Province of Los Ríos, Ecuador, on the premises of Mr. Gerardo Cepeda Alcívar.

The area where the present research was carried out presented the following meteorological characteristics on average:

Table 3. ***Weather conditions***

Parameters	
Temperature	24,9
Relative humidity	84,7
Altitude (masl)	120
Heliofania	
Annual precipitation (mm)	(mm)
Climate	Tropical humid
Topography	Plana

Source: INAMHI Meteorological Station, (INIAP 2015).

3.2. Materials

The material used to carry out this research is detailed in Table 4.

Table 4. Materials used for the establishment of a production of earthworm

Materials and equipment	Quantity
Rod (5m)	15
Pointed Cane (1.60m)	12
Flexometer	1
Weeding machine	1
Machete	2
Nails (2in)	5L
Sierra	1
Surgical Gloves	9
Zinc	12
Wire	1
Bins (20l)	6
Hammer	1
Black plastic (25M)	1
Buckets	2
Organic lime bag	30 L
Digital gramophone	1
Scissors	1
Watering cans	2
Humus	20kg

Paper paddle a4	1
Pencil	1
Printer	1
Computer	1
Calculator	1

3.3. Type of research

This research was of an explanatory type, using the observational method, as it focuses on describing the worm production in the different treatments.

3.3.1. Bibliographic or documentary method

It allowed to gather information on the topic of the research project and to compare it with the contribution given by various authors.

3.3.2. Deductive - inductive method

This method was used in the problem statement to determine the causes and effects of carrying out the research and it was possible to determine the causes of the development of a research in the incidence of the development of the research.

3.4. Preparation of the trial area

Chopped cane boxes and boards were used with dimensions of 60 cm width x 80 cm length and a height of 20 cm considering a major slope from ground level to the base of the box of 80 cm with a minor slope of 70 cm. The same dimensions were used in all treatments.

3.2.1. Features of compost bins

The compost bins are 40 cm wide, 60 cm long and 20 cm high, for a volume of 480 cm3 , to which 1.5 kg of red Californian earthworm was added.

3.5. Trial management

An area of 20 m^2 was used, in which a 2 m long branch was made to avoid direct sun rays to the compost bins and not to affect the earthworms, agricultural lime was placed on the soil, burnt oil was placed on the reeds to control the pests and the branch was covered with saran to avoid excess of sunlight on the earthworms and to avoid the entrance of pests.

From 45 days after the worms were placed, the reproduction of the worms was observed in each of the substrates, for which weekly samplings were carried out with a sample of 500 g of substrate in each experimental unit. The number of young worms, adult worms without clitellum and adult worms with clitellum and eggs were counted in the same sample.

Each of the different compost bins is moistened with 1 L of water before feeding each experimental unit.

In addition, 1.5 kg of humus was incorporated into the different compost bins and 1.5 kg of earthworms were placed in each of the treatments under study, to be fed with 3 kg of the different food substrates per experimental unit.

3.6. Treatments studied

The three treatments were studied and are detailed in the table below.

Table 5. Treatments studied

Treatments	
T1	Bovine manure 0.75 kg/mulch
T2	Organic waste 0.75 kg/compost 0.75 kg/compost
T3	Plantain rachis 0.75 kg/compost 0.75 kg/compost

3.7.1. Number of young earthworms

Young worms were counted 45 days after worm placement, from a 500 g sample taken at random from each of the compost bins corresponding to their treatment.

3.7.2. Number of adult worms without clitellum

Adult worms without clitellum were counted from a 500 g sample obtained from the corresponding treatments, noting the size of the worms and whether they had clitellum.

3.7.3. Number of adult worms with clitellum

The number of adult worms with clitellum was counted by means of a 500 g sample obtained from the corresponding treatments, evaluating those worms with the largest size and with the protrusion of their clitellum.

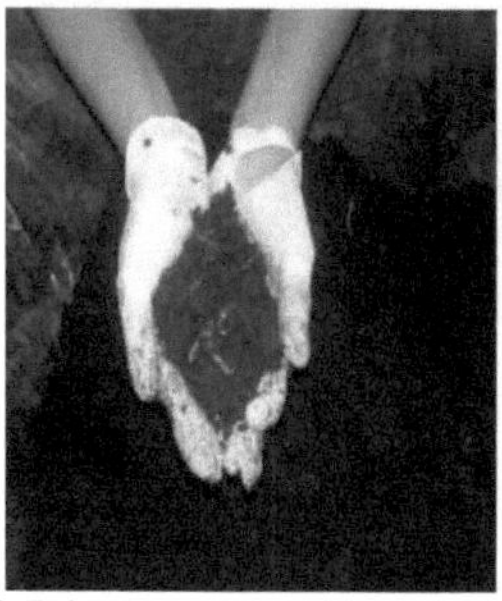

3.7.4. Number of eggs

The number of eggs was counted within the same samples taken in the treatments.

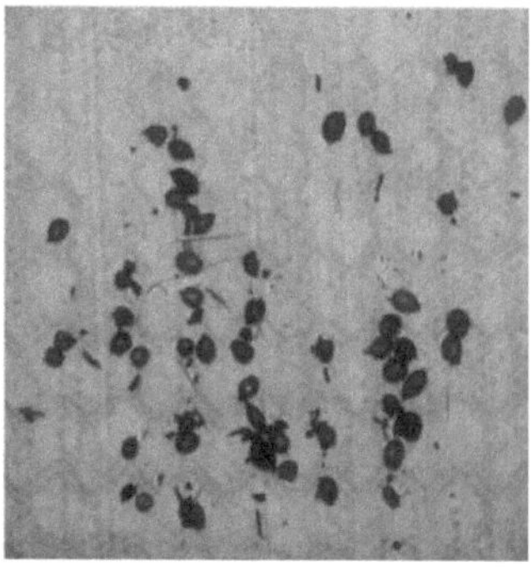

3.7.5. Lombrihumus increase

A screen was placed with a portion of feed according to the treatments to separate the worms from the humus and the worm castings were collected to be dried under ambient conditions and subsequently weighed.

3.8. Experimental design and statistical analysis

The experiment was conducted in a Completely Randomised Block Design (CBRD) with three treatments and four replicates for each treatment.

An analysis of variance (ANDEVA) was performed for the reproductive variables using the INFOSTAT statistical programme. Significant effects were subjected to a separation of means using Tukey's method with a significance level of ($P<0.05$).

CHAPTER IV

4.1. Results

4.1.1. Feeding of *E. foetida*

In the adult variable without clitellum there was a significant difference ($P<0.05$) between the treatments corresponding to the different substrates of feeding of the red Californian earthworm, being the treatment T2 (organic waste), the one that presented a higher number of adult earthworms without clitellum in comparison with the ones fed with banana rachis.

For the other variables analysed there is no significant difference, however, there is a numerical difference **(Table 1).**

Table 1. ***Variables analysed for the different feeding treatments of E. foetidae.***

Treatments	Variables					
	# Of worms NS	# Worms young people NS	# Adults without clitelium	# Adults with clitellum	Worm weight (gr) NS	# Little eggs NS
Bovine manure	157,25	83,54	51.77 ab	32,33	79,46	48,43
Organic waste	149,00	70,03	56,80 a	40,90	80,10	45,53
Banana rachis	140,80	75,73	37,88 b	23,75	72,03	45,13
AVERAGE	**149,02**	**76,43**	**48,82**	**32,33**	**77,20**	**46,36**
TUKEY 5% TUKEY 5% TUKEY 5%	**49,14**	**34,93**	**15,67**	**22,06**	**12,96**	**16,42**
CV	**15,20**	**21,07**	**14,80**	**31,45**	**7,74**	**16,32**

NS = Not significant

Table 2. ***Nutrient content of humus in the different treatments 8 months after the trial was established.***

Treatments	N	P	K	Ca	Mg	P^H	C.E (mmhos/cm)
	Concentration %					*Averages*	
TI Bovine dung	1,7	0,55	0,86	2,13	0,59	7,8	13,1
T2 Organic waste	1,6	0,57	1,39	1,86	0,50	8,0	23,7
T3 Banana rachis	1,8	0,57	1,60	1,90	0,51	8,0	12,9

The results of the chemical analysis of the earthworm humus are presented. In general, the highest concentration of primary macronutrients was found in T3 (banana rachis) with values of 1.8 (N) 0.57 (P) and 1.6 (K) compared to the concentration of bovine manure 1.7 (N) 0.55 (P) 0.86 (K) and organic waste 1.6 (N) 0.57 (P) 1.39 (K), However, with respect to the secondary macronutrients, a higher concentration of secondary nutrients can be seen in T1 (bovine manure) with values of 2.13 (Ca) and 0.59 (Mg) in comparison with the other treatments, which is corroborated with the research proposed by (García and Milán, 1994) in which the plantain rachis was better than the other treatments.

Table 3. ***Nutrient content of humus leachate in the different treatments 8 months after trial establishment***

Treatments	Conce ntration % %					Averages	
	N	P	K	Ca	Mg	pH	C.E(mmhos/cm)
TI Leachate Bovine manure	0,06	0,04	0,45	0,40	0,06	8,3	8,1
T2 Leachate Organic Waste	0,06	0,02	0,49	0,41	0,05	7,6	8,4
T3 Banana Rachis Leachate	0,06	0,02	0,40	0,41	0,05	5,6	8,3

The analysis of primary nutrient content in the earthworm leachate determined that the treatments TI, T2, T3 corresponding to the element of **N** were presented in equal concentration, in addition a difference in the values for **P** was observed in TI and **K** was higher than the other treatments**.** In the secondary elements analysed there is a minimum percentage of concentration in the different treatments, which agrees with the elements reported in the analyses mentioned by (Reines, 2008). It was observed that humus production at eight months was higher in solid in the TI, while in leached humus production was higher in the T3 **(Table 4).**

Table 4. ***Humus production and earthworm leachate production in the different***

treatments a

the 8 months of stablishment

Treatments	Humus production (kg)	Leachate production (L)
TI Bovine dung	19,5	8,75
T2 Organic waste	3,5	11,00
T3 Banana rachis	3,2	10,00

CHAPTER V.

Conclusions

- In the reproduction of worms with clitellum, number of eggs, number of worms, there was no significant difference, however, for adults without clitellum there was a difference of ($P<0.05$).

- Humus production was best with the cattle manure treatment, while leachate production was best with organic waste.

- The concentration of primary macronutrients was better with the plantain rachis treatment, and for secondary macronutrients the bovine manure.

showed better levels of nutrient concentration.

- The pH in the three treatments was slightly alkaline, while the electrical conductivity showed differences in T2 (organic waste).

- The concentration of nutrients in the leachate was similar in the different treatments. The pH in all three treatments was slightly alkaline.

Recommendations

- Use the bovine manure treatment if higher humus production is desired, and the bovine manure IT treatment if higher leachate volume is desired, considering macronutrient concentration, pH and EC.

- Conduct studies applying the humus produced in a field crop in solids and liquid.

- Conduct studies with different types of light intensity on the development and growth of earthworms.

CHAPTER VI

. Bibliography

Alastre, E. (2009). Lumbricutura. Mimeographed work from a course given at Universidad Central de Venezuela. Faculties of Veterinary Sciences and Agronomy.

Arledge, J. (2010). Increased yields using the red coquette worm. DIGESA.

Cando, M. (2010). The breeding of red earthworms. Quito, Ecuador.

Castells, X. (2012). Industrial waste recycling: Municipal solid waste and sewage sludge. Ediciones Díaz de Santos.

Dalzell, H. (2011). Soil management, production and use of compost in tropical and subtropical environments. Food & Agriculture Org.

De la Ossa, J. (2003). Guía para la cría, manejo y aprovechamiento sostenible de algunas especies animales promisorias y otras domésticas. Convenio Andrés Bello.

Del Pilar, C. (2013). Bases Químicas Del Medio Ambiente Unidad Didáctica. Editorial UNED.

Díaz, D. (2009). Growth and production dynamics of the Californian red earthworm (Eisenia foetida). ABA Foundation.

Dimas, N. R., Ríos, P. C., Viramontes, U. F., Gil, A. P., Chávez, E. F., Reyna, V. D. P. Á.,... & Reséndez, A. M. 2008. Tomato production in greenhouses with earthworm humus as substrate. Revista Fitotecnia Mexicana, 31(3), 265-272.

Domínguez, J., Lazcano, C., & Gómez-Brandón, M. 2010. Influence of vermicompost on plant growth: Contributions to the development of an objective concept. Acta zoológica mexicana, 26 (SPE. 2), 359-371.

Durán, L., & Henríquez, C. 2009. Growth and reproduction of the red earthworm (Eisenia foetida) in five organic substrates. Agronomía Costarricense, 33(2).

Esteve, D. (2007). Vermiculture as an alternative for the use of organic waste. Mexico : Universidad Autónoma de Tamaulipas.

Ferruzzi, C. (2009). Manual de lombricultura, . Madrid: Mundi - Press.

Flores-Pacheco, J.A. 2010. Vermiculture Management Manual. Technical guide. Futuro forestal. Managua, Nicaragua

García, D. (2009). Propiedades Fisicas de los Suelos. Bogota : Subdirección agrologica.

García, M. D., Oruña, L., Domínguez, H., & Martínez, V. 2005. Evaluation of the protein quality of earthworm (Eisenia foetida) meal in growing rats. Cuban Journal of Agricultural Science, 39(3).

Herrán, J., Torres, R. R. S., & Rojo, G. E. 2008. Importance of organic fertilizers. Ra Ximhai, 4(1), 57-67.

Martínez, E. (2012). Use of vermiculture. Application in the treatment of sludge from sewage treatment plants. Transport, Development and Environment.

Méndez Peña, H. 1.2015. Use of cocoa, oil palm and banana rachis harvest by-products in the production of humus with red Californian earthworm (Eisenia foetida) (Bachelor's thesis, Quevedo: UTEQ).

Peña, T. (2010). Manual for the production of organic fertilisers in urban production. Institute for Fundamental Research in Agriculture.

Reines, M. (2008). Earthworms with commercial value: biology and cultivation techniques. UQROO, Rodríguez, F. (2010). Lombricultura Para Pequeños Emprendedores. La Quimera.Stanley, E. (2006). Introduction to environmental chemistry. Reverte.

Tineo, B. (2010). Rearing and management of earthworms for agricultural purposes. Turrialba: CR. CATIE. Velásquez, J. (2003). Guía para la cría, manejo y aprovechamiento sostenible de algunas especies animales promisorias y otras domésticas. Madrid: algunas especies animales promisorias y otras domésticas.

Zabala, M. F., & Olate, E. 2003. Agronomic evaluation of humic substances derived from earthworm humus (Doctoral dissertation, Pontificia Universidad Católica de Chile).

Zenato , A. (2010). Alimentación Animal. Cuenca: Universidad del Azuay.

Table of contents

MIX
Papier aus verantwortungsvollen Quellen
Paper from responsible sources
FSC® C105338

Printed by Books on Demand GmbH, Norderstedt / Germany